ENGLISH ELECTRIC CLASS 50 DIESELS

FROM THE WESTERN REGION TO PRESERVATION

ENGLISH ELECTRIC CLASS 50 DIESELS

FROM THE WESTERN REGION TO PRESERVATION

PETER J. GREEN

PEN & SWORD TRANSPORT

AN IMPRINT OF PEN & SWORD BOOKS LTD.
YORKSHIRE – PHILADELPHIA

First published in Great Britain in 2022 by
Pen and Sword Transport
An imprint of
Pen & Sword Books Ltd
Yorkshire - Philadelphia

Copyright © Peter J. Green, 2022

ISBN 978 1 39901 782 4

The right of Peter J. Green to be identified as the author of this work has been asserted by him in accordance with the Copyright, Designs and Patents Act 1988.

A CIP catalogue record for this book is available from the British Library.

All rights reserved. No part of this book may be reproduced or transmitted in any form or by any means, electronic or mechanical including photocopying, recording or by any information storage and retrieval system, without permission from the Publisher in writing.

Typeset by SJmagic DESIGN SERVICES, India.

Printed and bound in India by Replika Press Pvt. Ltd.

Pen & Sword Books Ltd incorporates the Imprints of Pen & Sword Books Archaeology, Atlas, Aviation, Battleground, Discovery, Family History, History, Maritime, Military, Naval, Politics, Railways, Select, Transport, True Crime, Fiction, Frontline Books, Leo Cooper, Praetorian Press, Seaforth Publishing, Wharncliffe and White Owl.

For a complete list of Pen & Sword titles please contact

PEN & SWORD BOOKS LIMITED
47 Church Street, Barnsley, South Yorkshire, S70 2AS, England
E-mail: enquiries@pen-and-sword.co.uk
Website: www.pen-and-sword.co.uk

or

PEN AND SWORD BOOKS
1950 Lawrence Rd, Havertown, PA 19083, USA
E-mail: Uspen-and-sword@casematepublishers.com
Website: www.penandswordbooks.com

CONTENTS

Preface 6

Acknowledgements 8

Map 9

The Class 50: A First Encounter .... Photo 1

The Fifty Class 50s .... Photos 2-219

Class 50s on Preserved Railways .... Photos 220-227

Class 50s on Railtours .... Photos 228-260

Class 50s for Scrap .... Photos 261-263

Class 50s in Colour .... Photos 264-298

Class 50 Nameplates and Crests .... Photos 299-302

The Portuguese 1800 Class .... Photo 303

Bibliography 166

Index to Class 50 Locomotives by Photo Numbers 167

Index to Locations by Photo Numbers 168

PREFACE

In 1962, English Electric built the Type 4 diesel-electric prototype DP2 at the Vulcan Foundry at Newton-le-Willows. The locomotive was based on a Deltic body, but was fitted with a single English Electric 16CSVT engine of 2,700hp.

Initially British Railways decided to standardise on the Brush-Sulzer Type 4 design, later Class 47, but in 1965 the requirement for additional Type 4 locomotives was identified. Fifty locomotives were ordered, based on DP2, fitted with a British Railways Board standard cab design with a flat front and headcode box, as well as electronic control systems and various other internal modifications.

D400 entered service in October 1967 and the order was completed with D449 in November 1968, the locomotives initially being supplied on a ten-year lease. Fitted with the same 16CSVT engines as the prototype, they were designed to run at 100mph. In the TOPS renumbering scheme, they became the Class 50, and were nicknamed 'Hoovers' because of the sound made by the dynamic braking resistor cooling fans.

Initially they were used on passenger services on the non-electrified section of the West Coast Main Line, north of Crewe. From May 1970 they were used in pairs, working in multiple, on certain services to Scotland, allowing schedules to be reduced. By 1974 further electrification of the West Coast Main Line allowed the class to be transferred to the Western Region, where they worked passenger services out of Paddington, replacing the Western Class diesel-hydraulics. In the late 1970s the locomotives were named after ships of the Royal Navy, with 50035 being the first, receiving the name *Ark Royal* in January 1978.

Following the introduction of the InterCity 125 units, from 1977 the Class 50s were transferred to passenger services on other routes, including the Waterloo to Exeter line. Because of increasing reliability problems, the class was refurbished at Doncaster Works between 1979 and 1984. The work included removing much of the equipment requested by the British Railways Board. The electronics were simplified, the air intake fan arrangement was modified, and the slow speed control and rheostatic braking were removed. The locomotives were also fitted with high intensity headlights. Following refurbishment, all the class were based at Old Oak Common and Plymouth Laira depots and, from 1980, were repainted in Large Logo livery.

Various other liveries were applied to members of the class, including 50007 which was painted in lined Brunswick Green and renamed *Sir Edward Elgar* for the 150th anniversary of the Great Western Railway in 1984. From 1986, many of the class were repainted in Network SouthEast livery and worked trains within that passenger sector from Paddington and Waterloo. In the late 1980s, some locomotives were transferred to the civil engineers' department and worked on permanent way trains, while 50049 was modified to work on freight trains, renumbered 50149 and painted in two-tone grey. It was later restored to its earlier condition and identity.

Withdrawals commenced in 1987 with 50011 *Centurion*. Three locomotives, 50007, 033 and 050, were retained for railtours until the final withdrawals took place in March 1994. Eighteen locomotives have survived into preservation, some of which are certified to run on the main line.

Ten similar broad-gauge locomotives, fitted with the same 16CSVT engines, were also supplied to Portugal in 1968 and 1969. All were withdrawn by 2001.

Over the years, photographing and travelling behind the Class 50 diesels has given me a considerable amount of pleasure, and with all the locomotives that are still in operation, both on the main line and on preserved railways, hopefully this will continue for some years to come.

Peter J. Green
Worcester, England,
December 2020

ACKNOWLEDGEMENTS

I would like to thank Paul Dorney for contributing a number of his photographs, which have helped to fill a few of the gaps in this book, and for assisting with various captions for my own photographs.

Thanks are also due to Steve Turner, James Billingham and Martin Loader for their assistance with some captions, where my notes have let me down. Martin's excellent website, www.hondawanderer.com, has also proved useful for extra detail in a number of instances. It is surprising how often we chose the same location to photograph special trains!

Once again, Val Brown has checked my text for errors. It is very much appreciated.

MAP • 9

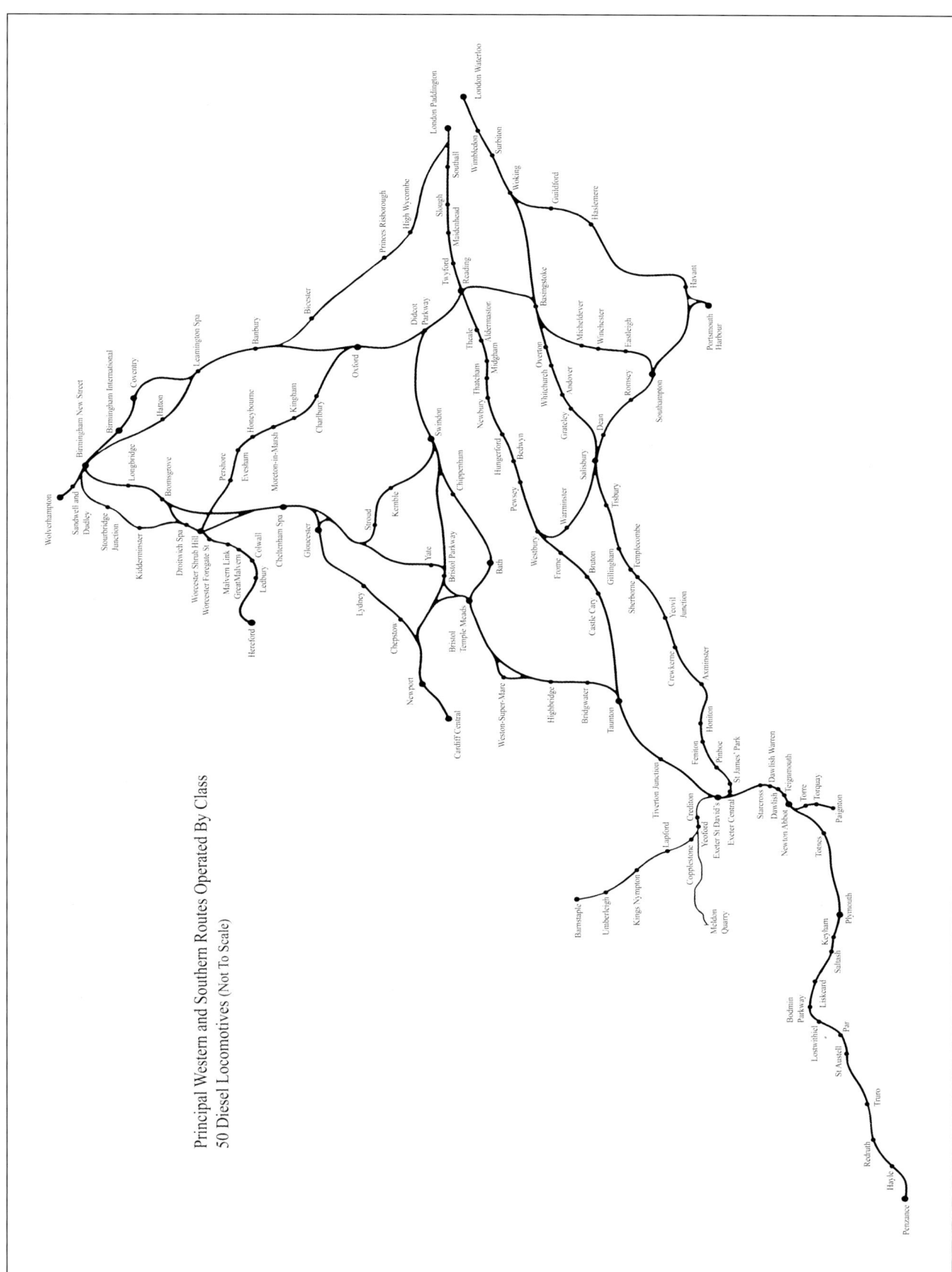

Principal Western and Southern Routes Operated By Class 50 Diesel Locomotives (Not To Scale)

This map shows the principal routes over which the Class 50 locomotives hauled trains during their time in normal service on the former Great Western and Southern Railway lines of British Railways. Many other lines in the area covered have been omitted for clarity.

THE CLASS 50: A FIRST ENCOUNTER

1. In 1975, I visited Devon to photograph some of the remaining Western Class diesel-hydraulics that were still operating in the area. At that time, Class 50 diesels transferred from the London Midland Region were slowly replacing them on front line duties. Here, an unidentified Class 50, as yet unnamed, is pictured heading along the sea wall at Dawlish, with the 1A05 12.00 Paignton to Paddington. This was the first time I had seen these locomotives in action, which I have to admit I was less than pleased about, since I was there to photograph the Westerns. It was not until the early 1980s, when they were named and painted in Large Logo livery, that my interest in the class really took off, an interest that has continued to the present day. 31 March 1975.

THE FIFTY CLASS 50s

2. 50001 *Dreadnought*. The 07.35 Cardiff Central to Glasgow Central arrives at the summit of the Lickey Incline, at Blackwell, behind 50001 *Dreadnought* and 50042 *Triumph*, both in Large Logo livery. This working often had a pair of Class 50s in charge, as far as Birmingham New Street, at this time. 21 March 1987.

3. 50001 *Dreadnought*. In the original livery for the class, *Dreadnought* waits for its departure time at Worcester Shrub Hill station with a train to Paddington. 22 November 1981.

4. 50001 *Dreadnought*. Previously numbered D401, *Dreadnought* stands at Plymouth. This locomotive was the second member of the class to enter service in December 1967. 28 August 1983.

5. 50001 *Dreadnought*. With electric multiple unit sheds in the background, 50001 *Dreadnought* approaches Gap Road bridge at Wimbledon with the 13.15 Waterloo to Exeter St David's. 1 August 1989.

6. 50001 *Dreadnought*. Viewed from Bishopstoke Road bridge, 50001 *Dreadnought* approaches Eastleigh with the Sundays only 10.00 Waterloo to Salisbury. The freight line to Romsey, diverging to the left, reopened to passengers in 2003. 12 August 1990.

7. **50002 *Superb*.** The 9M01 Tuesdays only civil engineers' train from Gloucester to Bescot passes Wednesbury with *Superb* in charge. Wednesbury No. 1 signal box and the embankment of the old Great Western line are in the background. The line through Wednesbury closed in March 1993. 14 June 1988. *(Photo courtesy Paul Dorney)*

8. 50002 *Superb*. With the Malvern Hills behind, 50002 *Superb* approaches Newland level crossing, between Malvern Link and Worcester, with the 18.15 Hereford to Paddington. The area on the right is the site of the former Newland permanent way depot. 27 April 1986.

9. 50002 *Superb*. The old goods shed at Chard Junction is still standing as *Superb* passes with the 08.11 Exeter St David's to Waterloo. 20 April 1991.

10. 50002 *Superb*. In revised Network SouthEast livery, *Superb* passes Exmouth Junction Coal Concentration Depot with the 11.15 Waterloo to Exeter St David's. 31 July 1990.

11. 50002 *Superb*. Dinton station was closed to passenger traffic in 1966 and to goods traffic the following year. Here, *Superb* passes the old station with the 10.20 Exeter St David's to Waterloo. 22 June 1991.

12. 50002 *Superb*. The 6M73 Quidhampton to Willesden china clay slurry tanks runs through Grateley station behind *Superb* and Brush Type 4 47052. 17 August 1991.

13. 50002 *Superb*. Viewed from the Newbury Road bridge, *Superb* passes Whitchurch with the 08.10 Exeter St David's to Waterloo. 20 July 1991.

14. 50002 *Superb*. In Large Logo livery, 50002 *Superb* arrives at Dawlish with an Exeter St David's to Paignton working. Dawlish signal box, located on the up platform, was demolished in 2013. 6 October 1984.

15. 50003 *Temeraire*. With the New North Road bridge in the background, 50003 *Temeraire* and 50001 *Dreadnought* pass Exeter Central signal box as they depart from Exeter Central station with the 08.11 Exeter St David's to Waterloo. The signal box, opened in 1927, is a Southern Railway design, but with a hipped roof. It was closed in 1985. 31 August 1989.

16. 50003 *Temeraire*. After replacing an electric locomotive at Birmingham New Street, *Temeraire* rounds the curve on the approach to Abbotswood Junction, near Worcester, with the 13.11 Manchester Piccadilly to Paignton. This location is now the site of Worcestershire Parkway station. 19 May 1984.

17. 50003 *Temeraire*. On Sundays from 18 June to 9 July in 1989, Waterloo to Exeter trains were routed through Southampton. Here, *Temeraire* heads the diverted 14.40 Waterloo to Exeter St David's away from Southampton station. 25 June 1989.

18. 50004 *St Vincent*. Running alongside the River Exe estuary, *St Vincent* passes the well-known location of Cockwood Harbour, between Dawlish Warren and Starcross, with the 17.35 Paignton to Waterloo. 27 August 1989.

19. 50004 *St Vincent*. With 50004 *St Vincent* in charge, an unidentified northbound working approaches Oddingley level crossing, between Abbotswood Junction and Stoke Works Junction. 2 May 1984.

20. 50004 *St Vincent*. Class 50s did not normally run through Droitwich Spa until they were relegated to engineers' trains but, in 1982, there was a regular northbound empty stock working from Worcester to Birmingham New Street. Here, *St Vincent* is pictured heading the empty stock away from Droitwich Spa station. 1 March 1982.

21. 50004 *St Vincent*. Heading the Sundays only 16.15 Hereford to Paddington, *St Vincent* arrives at Kingham station. The embankment of the former Banbury to Cheltenham line can be seen in the background. Connections between this line and Kingham station formed a triangle on which locomotives could be turned. Note the redundant platform faces for the Chipping Norton and Banbury trains on the right. 10 April 1988.

22. 50005 *Collingwood*. With the River Plym on the right, *Collingwood* heads the empty stock of the 14.42 Plymouth to Waterloo from Laira to Plymouth station. 27 August 1989.

23. 50005 *Collingwood*. Tisbury Gates, a level crossing over a minor road west of Tisbury station, was a popular location for photographing the Waterloo to Exeter trains. Here, 50005 *Collingwood* passes the crossing with the 09.28 Exeter St David's to Waterloo. 20 August 1989.

24. **50005 *Collingwood*.** Exiting Parson's Tunnel, *Collingwood* heads along the sea wall towards Teignmouth with the 08.15 Basingstoke to Paignton. 29 July 1990.

25. 50005 *Collingwood*. With the City of Worcester in the distance, 50005 *Collingwood* passes Henwick with the 16.10 Paddington to Hereford. This is the site of the former Henwick station, closed in 1965. The signal box is next to the rear of the train. 2 June 1985.

26. 50005 *Collingwood*. Permanent way works provide a convenient pile of ballast to stand on for a photograph of *Collingwood* heading south near Dunhampstead, between Stoke Works Junction and Abbotswood Junction, with the 1V90 13.20 Liverpool Lime Street to Plymouth. 5 May 1984.

27. 50006 *Neptune*. After arriving with empty stock and running round, 50006 *Neptune* stands in Midland Yard, Worcester Shrub Hill. It will later run to Cheltenham Spa before working to Paddington. 20 July 1986.

28. 50006 *Neptune*. Working a Paignton to Newton Abbot service, *Neptune* passes Aller Junction, near Newton Abbot. The line to Plymouth diverges to the right. In 1987, the junction was moved closer to Newton Abbot station. 21 September 1985.

29. 50007 *Sir Edward Elgar*. In 1984, for the 150th anniversary of the Great Western Railway, 50007 *Hercules* was renamed *Sir Edward Elgar* and painted into lined Brunswick Green livery. Here, 50007 heads away from Eastleigh with the diverted 10.00 Waterloo to Salisbury, on a fine Sunday in June 1989. 25 June 1989.

30. 50007 *Sir Edward Elgar*. Approaching Redbridge, near Southampton, *Sir Edward Elgar* heads the diverted 14.28 Exeter St David's to Waterloo past the old Redbridge East Bridge over the River Test. The bridge is a Grade II listed structure. 12 August 1990.

31. 50007 *Sir Edward Elgar*. Pictured on permanent way duty, *Sir Edward Elgar* stands at Southall, while its train is loaded with old ballast during the remodelling of Southall East Junction. 4 February 1989.

32. 50007 *Sir Edward Elgar*. The last day of the Class 50s on the Waterloo to Exeter trains saw 50007 *Sir Edward Elgar* and 50050 *Fearless*, carrying its original D400 number and without nameplates, in action. Here, the pair pass Milborne Wick, between Templecombe and Sherborne, with the 16.55 Waterloo to Exeter St David's. The headboard reads 'Farewell Class 50'. 24 May 1992.

33. 50008 *Thunderer*. Before the semaphore signals were replaced in 1986, *Thunderer* approaches Silk Mill crossing, west of Taunton station, with the 13.44 Birmingham New Street to Paignton. 14 July 1984.

34. 50008 *Thunderer*. With a loaded ballast train from the British Rail ballast quarry at Meldon, *Thunderer* runs through Okehampton station. Passenger services at Okehampton ceased in 1972, recommencing in 1997. 1 July 1990.

35. 50008 *Thunderer*. With the A30 Okehampton Bypass in the background, 50008 *Thunderer* heads east with a loaded ballast train from Meldon Quarry. The quarry is three miles from Okehampton. 3 August 1990.

36. 50008 *Thunderer*. In some fine winter sunshine, 50008 *Thunderer* heads away from Worcester Shrub Hill with the diverted 1V76 09.20 Liverpool Lime Street to Penzance. 31 January 1984.

37. 50008 *Thunderer*. With *Thunderer* in charge, the 07.30 Penzance to Aberdeen pauses at Par, the junction of the line to Newquay. A Class 37 waits with a china clay train on the other side of the platform. A second Class 37 is alongside it, with an inspection saloon behind. 31 August 1984.

38. 50009 *Conqueror*. Penzance is the terminus of the former Great Western Railway in Cornwall, 327 miles from London Paddington via Bristol Temple Meads. Here, *Conqueror* makes a smoky departure from Penzance Platform 3 with the 14.15 to Waterloo. Other Class 50s and a Class 47 are in the sidings on the left and an InterCity 125 unit stands in Platform 1. 26 August 1984.

39. 50009 *Conqueror*. The lines from Worcester to Oxford and Cheltenham Spa divide at Norton Junction. Here, *Conqueror* approaches Norton Junction, near Worcester, with a loaded ballast train from Bescot Yard to Gloucester. 28 February 1989.

40. 50009 *Conqueror*. A second view of 50009 *Conqueror* with the Bescot to Gloucester ballast. Here, the train is pictured heading away from Abbotswood Junction towards Cheltenham Spa and Gloucester. Abbotswood Junction is where the line from Norton Junction joins the Birmingham to Bristol main line. 28 February 1989.

41. 50009 *Conqueror*. With the moorings on the River Hamble in the foreground, *Conqueror* crosses the river at Bursledon with the 09.05 Brighton to Plymouth. 11 August 1990.

42. 50009 *Conqueror*. On a sunny evening, 50009 *Conqueror* crosses the River Severn, at Worcester, with the 1C56 17.00 Paddington to Hereford. 23 April 1982.

43. 50010 *Monarch*. Waiting for departure time, *Monarch* stands at Plymouth with the Sundays only 1V84 10.46 Liverpool Lime Street to Penzance. 28 August 1983.

44. 50010 *Monarch*. With snow on the ground and scaffolding around the bridge over the Worcester and Birmingham Canal, *Monarch* approaches Worcester Foregate Street station, with a Paddington to Hereford service. 10 January 1982.

45. 50011 *Centurion*. The first Class 50 to be withdrawn, in February 1987, was 50011 *Centurion*. The locomotive is pictured here running through Dawlish station with the 14.45 Paddington to Plymouth. 15 September 1985.

46. 50011 *Centurion*. With 50011 in charge, the 2B10 08.25 Plymouth to Penzance approaches Liskeard. The line curving away to the left, opposite the Great Western signal box, joins the Looe to Moorswater line at Coombe. 1 September 1984.

47. 50012 *Benbow*. The Lickey Incline, between Bromsgrove and Blackwell, is just over two miles long, with an average gradient of 1 in 37.7. Here, *Benbow* climbs the bank at Pikes Pool Lane, Burcot, with the 1S39 07.50 Paignton to Glasgow Central. An electric locomotive will take over the train at Birmingham New Street. 5 May 1984.

48. 50012 *Benbow*. Pictured between Barnt Green and Blackwell, 50012 *Benbow* heads south from Birmingham New Street, with the 1V85 11.33 Manchester Piccadilly to Plymouth. In the background, construction of the M42 motorway bridge over the railway is progressing. 5 May 1984.

49. 50013 *Agincourt*. The fourth member of the class to be withdrawn, in April 1988, 50013 *Agincourt* shunts passenger stock past the fine bracket signal at the south end of Exeter St David's station. 16 March 1985.

50. 50014 *Warspite*. The up 'Torbay Express', the 11.05 Paignton to Paddington, arrives at Newton Abbot behind 50014 *Warspite*. 6 August 1985. *(Photo courtesy Paul Dorney)*

51. 50014 *Warspite*. Passing Blackwell, *Warspite* arrives at the summit of the Lickey Incline, with the 1M22 11.38 Plymouth to Manchester Piccadilly. 15 December 1983.

52. 50014 *Warspite*. Before receiving its name in May 1978, 50014 stands in the sun at Bristol Temple Meads station. September 1976. *(Photo courtesy Paul Dorney)*

53. 50015 Valiant. With a loaded ballast train from Meldon Quarry, *Valiant* approaches Crediton signal box, located next to the level crossing at the west end of the station. 30 August 1989.

54. 50015 Valiant. Passenger trains only stop at Yeoford, between Crediton and Copplestone, on request. Here, *Valiant* runs past the old yard with a ballast train from Meldon Quarry. 29 August 1989.

55. 50015 Valiant. Crossing the River Usk and approaching Maindee West Junction at Newport, *Valiant* heads the 3A12 06.44 Milford Haven to Old Oak Common newspaper vans. Note the inspection saloon and TPO behind the locomotive. The bridge was built in 1888, replacing an earlier bridge, and was widened to four tracks by 1911. Newport Castle is on the far left. 5 March 1988.

56. 50015 Valiant. Heading past Norton Junction signal box, *Valiant* takes the line to Abbotswood Junction with the 9B21 Worcester to Gloucester departmental working. 23 February 1989.

57. 50015 Valiant. Worcester Foregate Street station, located in the centre of Worcester, was opened by the Great Western Railway in 1860. Here, *Valiant* passes some of Worcester's semaphore signals as it arrives at the station with the Sundays only 16.10 Paddington to Hereford. 31 August 1986.

58. 50016 Barham. Heading a ballast train off the Kidderminster line, *Barham* approaches Droitwich Spa station. On the right is Ruston 48DS 0-4-0 shunter, *The Sheriff* (458961 of 1962), standing at the closed Underwood's coal depot. I worked at Droitwich for many years and was fortunate that my office overlooked this railway scene. 19 January 1989.

59. 50016 *Barham*. The 14.42 Plymouth to Waterloo passes Exminster signal box with 50016 *Barham* and 50017 *Royal Oak* in charge. The signal box, taken out of service in 1986, has since been removed for preservation. 3 September 1989.

60. 50017 *Royal Oak*. The 13.11 Manchester Piccadilly to Paignton passes Tiverton Junction station behind 50017 *Royal Oak*. The station closed in 1986 when Tiverton Parkway was opened on the site of the former Sampford Peverell station. 18 August 1984.

61. 50017 *Royal Oak*. Fenny Bridges is located in attractive countryside between Honiton and Feniton. Here, *Royal Oak* passes with the 13.15 Waterloo to Exeter St David's. 20 April 1991.

62. 50017 *Royal Oak*. The 08.11 Exeter St David's to Waterloo passes New Malden with 50017 *Royal Oak* in charge. 31 March 1989.

63. 50017 *Royal Oak.* Horse Cove, between Dawlish and Teignmouth, was the location of a number of Great Western Railway publicity photographs. Here, *Royal Oak* passes with the 11.14 Southampton to Plymouth. Dawlish, partly obscured by the mist, is visible in the background. 27 April 1991.

64. 50017 *Royal Oak.* A Great Western backing signal, complete with route indicator, stands on the up platform at Castle Cary station. The white-painted signal box, on the right, is a replacement for an earlier one, destroyed in 1942. Here, *Royal Oak* approaches the station with the 1A42 09.45 Paignton to Paddington. 14 July 1984.

65. 50017 *Royal Oak*. Malvern Link and Great Malvern are the two remaining stations serving Malvern. Here, *Royal Oak* pauses at Malvern Link station, located north of Great Malvern, with the Sundays only 16.15 Hereford to Paddington. 4 September 1988.

66. 50018 *Resolution*. The 11.15 Waterloo to Exeter St David's passes the old goods shed at Chard Junction, between Axminster and Crewkerne, behind 50018 *Resolution*. The station, previously the junction for the branch line to Chard Town, was closed in 1966. 13 April 1991.

67. 50018 *Resolution*. Passing the Grade II listed main station building, *Resolution* arrives at Crewkerne with the 16.22 Exeter St David's to Waterloo. *Resolution* was one of the few Class 50s fitted with ploughs at the time. 13 July 1991.

68. 50019 *Ramillies*. After painting into Laira Blue livery in 1989, 50019 *Ramillies* was displayed at the Railfreight Coal open day at Barry in 1990. English Electric Type 3 37350 is behind. *Ramillies* was withdrawn one month later, in September 1990. 19 August 1990.

69. 50019 *Ramillies*. In Laira Blue livery, *Ramillies* stands in front of two Class 47s at Exeter depot, next to St David's station. 30 July 1990.

70. 50020 *Revenge*. Bescot Yard, near Walsall, is a major freight yard in the West Midlands. Here, 50020 *Revenge* heads the Tuesdays only Bescot to Gloucester permanent way train away from the yard. 1 March 1990. *(Photo courtesy Paul Dorney)*

71. 50020 *Revenge*. With the tracks and signalling of the Severn Valley Railway on the right, *Revenge*, on ballast duty, crosses to the up line at Kidderminster. The semaphore signalling on the main line has since been replaced. 8 February 1989.

72. 50021 *Rodney*. Named after the battleship HMS *Rodney*, 50021 shunts the yard at Worcester Shrub Hill. This locomotive was withdrawn one year later and is now preserved. 16 February 1989.

73. 50021 *Rodney*. On permanent way duty associated with Sunday engineering work, *Rodney* stands in the yard behind Worcester Shrub Hill station at the head of a train of concrete sleepers. 5 March 1989.

74. 50021 *Rodney*. Bodmin Parkway station serves the town of Bodmin, situated three miles away. Here, 50021 *Rodney* heads round the curve, as it departs from the station with the 10.00 Penzance to Liverpool Lime Street. 25 August 1984.

75. **50021** *Rodney*. Soon after passing Bromsgrove, *Rodney* approaches Stoke Works Junction with the 1V85 11.33 Manchester Piccadilly to Plymouth. 18 January 1984.

76. **50022** *Anson*. The Royal Albert Bridge over the River Tamar at Saltash, designed by Isambard Kingdom Brunel, was opened in 1859. Here, *Anson* heads away from the bridge with an unidentified eastbound working. The Tamar Bridge, carrying the A38 trunk road, is in the background. 24 August 1984.

77. 50022 *Anson*. Passing Worcester Shrub Hill station signal box, *Anson* runs on to the 11.46 to Paddington. 15 April 1982.

78. 50022 *Anson*. With a parcels van behind the locomotive, 50022 *Anson* departs from Moreton-in-Marsh with the 11.46 Worcester Shrub Hill to Paddington. 15 April 1982.

79. 50023 *Howe*. Painted in the original Network SouthEast livery with raised stripes at each end, *Howe* departs from Didcot Parkway station with the 16.07 Paddington to Oxford. Compare this livery to that carried by *Howe* one year later, in the following photograph. 22 April 1989.

80. 50023 *Howe*. With Didcot Power Station in the background, 50023 *Howe*, in revised Network SouthEast livery, approaches Culham station with the 09.15 Paddington to Manchester Piccadilly. 7 April 1990.

81. 50023 *Howe*. Soon after leaving Paddington, *Howe* passes Subway Junction with the 15.15 Paddington to Oxford. The Metropolitan line of the London Underground, between Westbourne Park and Royal Oak, passes under the former Great Western line here. After July 1990, this section of the Metropolitan line became the Hammersmith & City line on the London Underground map. 31 March 1990.

82. 50023 *Howe*. With Southall water tower and gasometers in the background, 50023 *Howe* and 47598 approach Southall with the 12.00 Oxford to Paddington. *Howe* is in the early Network SouthEast livery, while the Class 47 carries the revised version. 50007 *Sir Edward Elgar* is on permanent way duty on the left. 4 February 1989.

83. 50023 *Howe*. Weybridge station is just over nineteen miles from London Waterloo. Here, *Howe* takes the middle road through the station with the 11.10 Waterloo to Exeter St David's. 10 May 1989.

84. 50023 *Howe*. Cheltenham Spa station, opened as Lansdown station by the Birmingham and Gloucester railway, was renamed Cheltenham Spa (Lansdown) in 1925 and Cheltenham Spa in 1977. Here, *Howe* arrives at the station with the 1V76 09.20 Liverpool Lime Street to Penzance. 16 October 1982.

85. 50023 *Howe*. Viewed from Langstone Rock, 50023 *Howe* rounds the curve between Dawlish and Dawlish Warren with the 08.20 InterCity Holidaymaker from Paignton to Glasgow Central. 26 August 1989.

86. 50024 *Vanguard*. Clapham Junction station, approximately two and three-quarter miles from London Victoria and four miles from London Waterloo, is one of the busiest in Europe. Here, *Vanguard* heads the 10.15 Waterloo to Salisbury soon after passing the station. 21 July 1990.

87. 50024 *Vanguard*. A DMU heads west, as the 14.40 Newbury to Paddington passes Aldermaston behind 50024 *Vanguard*. The siding, on the left, was made redundant by the construction of a new road bridge in 2012. 18 March 1989.

88. 50024 *Vanguard*. Crossing the River Blythe, *Vanguard* heads towards Hampton-in-Arden station with the 15.13 Paddington to Wolverhampton. 8 April 1990.

89. 50024 Vanguard. Heading for Leamington Spa, *Vanguard* runs through Harbury cutting with the 11.18 York to Paddington. 24 March 1990.

90. 50024 Vanguard. With just under eleven miles to go to London Paddington, *Vanguard* approaches Hayes and Harlington station with the 11.00 from Oxford. 18 March 1989.

91. 50024 Vanguard. Aynho Junction, five miles south of Banbury, is the junction of the lines to Banbury, Oxford and Bicester North. Here, Vanguard passes with the 07.05 Paddington to Wolverhampton. The up Chiltern line to Bicester North branches off to the left, while the down line is to the right of the signal box. The box became redundant in 1992 and was demolished around 2001. 12 August 1989.

92. 50024 Vanguard. The Sundays only 16.15 Hereford to Paddington departs from Worcester Shrub Hill behind 50024 Vanguard. 15 March 1987.

93. 50024 *Vanguard*. With *Vanguard* in charge, the 09.15 Paddington to Manchester Piccadilly is pictured near Kenilworth, heading for Coventry and Birmingham. 24 March 1990.

94. 50025 *Invincible*. Passing the old GWR station buildings, 50025 *Invincible* runs through Hanwell, between West Ealing and Southall, with the 14.15 Paddington to Oxford. The station was named Hanwell and Elthorne until 1974, as can be seen on the old station sign on the left. 4 February 1989.

95. 50025 *Invincible*. Before it was named in June 1978, 50025 heads the 1B35 10.30 Paddington to Paignton round the curve between Dawlish Warren and Dawlish. 31 March 1975.

96. 50025 *Invincible*. Before resignalling in late 1984 and early 1985, *Invincible* heads a Paddington to Paignton service through Castle Cary. 14 July 1984.

97. 50026 *Indomitable*. Leamington Spa is the junction of the lines to Banbury, Birmingham and Coventry. Here, 50026 *Indomitable* pauses at the station with the 09.15 Paddington to Manchester Piccadilly. 28 December 1989.

98. 50026 *Indomitable*. The 11.18 Wolverhampton to Paddington passes Greaves Siding signal box, opened in 1918, behind 50026 *Indomitable*. The signal box, located on the opposite side of the line to the former Harbury Cement Works, was a replacement for an earlier box. 6 September 1989.

99. 50026 *Indomitable*. Adderley Park station, between Birmingham New Street and Stechford, was opened in 1860 by the London and North Western Railway. Here, *Indomitable* passes with the 11.18 York to Paddington. 27 April 1990. *(Photo courtesy Paul Dorney)*

100. 50026 *Indomitable*. Diverted because of Sunday engineering work, *Indomitable* passes Stourbridge Junction Middle signal box with the 1V84 11.21 Liverpool Lime Street to Penzance. The signal box, a type 7B GWR design, was built in 1901. 12 February 1984.

101. 50026 *Indomitable*. Old Oak Common Traction Maintenance Depot, situated to the west of London, provided the motive power for Paddington station. With a Class 47 and a Class 31 in the background, *Indomitable* runs on to the turntable at the depot. 29 April 1990.

102. 50026 *Indomitable*. Still in BR Blue livery, *Indomitable* passes Norton Junction signal box, near Worcester, with the 1C40 15.00 Paddington to Hereford. 7 April 1982.

103. 50026 *Indomitable*. The brick-built Ledbury Viaduct, completed in 1860, has thirty-one spans. The bricks were made on site from clay removed during excavations for the foundations. Here, *Indomitable* crosses the Grade II listed viaduct as it heads west with an empty stock working from Worcester to Hereford. 27 September 1987.

104. 50027 *Lion*. On Easter Monday 1989, 50027 *Lion* departs from Basingstoke, forty-eight miles from Waterloo, with an unidentified working to Waterloo. 27 March 1989.

105. 50027 *Lion*. With Eastleigh Works in the background, 50027 *Lion* approaches Eastleigh station with Sunday's diverted 15.48 Exeter St David's to Waterloo. Originally opened in 1891 by the London and South Western Railway as a carriage and wagon works, Eastleigh later became the principal works of the Southern Railway. 25 June 1989.

106. 50027 *Lion*. Headed by *Lion*, a line of locomotives stands on the middle road at Worcester Shrub Hill for Worcester Rail Day on 22 May. Behind 50027 are 85101, 83012, D120, and 31116. The locomotives had arrived from Exeter after an open day there on 2 May. *Lion*, withdrawn in July 1991, was preserved and based on the North Yorkshire Moors Railway at this time. 11 May 1994.

107. 50027 *Lion*. Diverted from its usual route because of Sunday engineering work, 50027 *Lion* passes Great Wishford, as it heads south between Warminster and Salisbury, with the 09.05 Exeter St David's to Waterloo. 2 December 1990.

108. 50028 *Tiger*. Running alongside the River Teign estuary, 50028 *Tiger* and 50003 *Temeraire* make a fine sight as they approach Shaldon Bridge, Teignmouth, with the 11.25 Paignton to Waterloo. 27 August 1989.

109. 50028 *Tiger*. Malvern Wells railway station, on the Great Western line at Lower Wyche between Great Malvern and Colwall, closed in 1965. *Tiger* passes the site of the old station with the 16.15 Hereford to Paddington. Malvern Wells signal box is on the left. There was also a Malvern Wells station on the line to Upton-on-Severn which later became known as Malvern Hanley Road. 6 April 1986.

110. 50028 *Tiger*. The 1S61 07.35 Cardiff Central to Glasgow Central and Edinburgh, 'The Principality', passes Spetchley loop, between Abbotswood Junction and Stoke Works Junction, behind 50028 *Tiger*. This is the site of the former Spetchley station. Opened in 1840, it closed in 1855. Goods traffic ceased in January 1961. 9 May 1987.

111. 50028 *Tiger*. Heading away from Totnes station, *Tiger* begins the climb of Rattery Bank with the 09.05 Brighton to Plymouth. Rattery Bank, one of the South Devon Banks between Exeter and Plymouth, commences at Totnes and reaches the summit at Wrangaton. The railway then descends Hemerdon Bank to Plympton, near Plymouth. 26 August 1989.

112. 50028 *Tiger*. Pictured between Warminster and Salisbury, *Tiger* heads Sunday's diverted 11.55 Exeter St David's to Waterloo past Little Langford. 25 November 1990.

113. 50028 *Tiger*. Sunday's diverted 14.00 Waterloo to Exeter St David's passes St Denys, near Southampton, with 50028 *Tiger* in charge. 23 September 1990.

114. 50028 *Tiger*. London's Waterloo station, opened in 1848 by the London and South Western Railway, is the terminus of the West of England services to Exeter, as well as other main line and commuter services. Here, *Tiger* departs from the station with the 17.10 to Exeter St David's. 15 April 1989.

115. 50029 *Renown*. In the original Network SouthEast livery, 50029 *Renown* arrives at Crewkerne with the 1V09 09.10 Waterloo to Exeter St David's. 19 November 1988.

116. 50029 *Renown*. In revised Network SouthEast livery, *Renown* heads the 06.45 Exeter St David's to Waterloo near Pirbright, between Farnborough and Brookwood. 11 May 1991.

117. 50029 *Renown*. With a diesel-electric multiple unit, forming the 13.24 Salisbury to Basingstoke, on the left, 50029 *Renown* waits for its next duty at Salisbury. 20 August 1989.

118. 50029 *Renown*. Passing Wimbledon, *Renown* heads a 4TC set on the 08.09 Salisbury to Waterloo. The Trailer Control (TC) units were formations of three or four carriages with a driving position at each end of the set. 1 August 1989.

119. 50029 *Renown*. Rounding the curve on the Plymouth line, *Renown* heads away from Newton Abbot with the 08.55 Waterloo to Exeter St David's, extended to Plymouth. 15 September 1991.

120. 50030 *Repulse*. On a rainy day, passengers wait under the station canopy as 50030 *Repulse* arrives at Sherborne with the 09.40 Plymouth to Brighton. 28 October 1989.

121. 50030 *Repulse*. Heading away from Yeovil Junction, *Repulse* is about to pass under the A37 road bridge with the 07.45 Basingstoke to Exeter St David's. 2 November 1991.

122. 50030 *Repulse*. Running alongside the Worcester and Birmingham Canal, 50030 *Repulse* and 50009 *Conqueror* head the 1S61 07.35 Cardiff Central to Glasgow Central, near Birmingham University. 28 March 1987.

123. 50030 *Repulse*. After passing Barnt Green, the 1V76 09.20 Liverpool Lime Street to Penzance approaches Blackwell and heads for Bromsgrove, behind 50030 *Repulse*. 27 December 1983.

124. 50030 *Repulse*. After leaving Ledbury Tunnel, *Repulse* heads for Colwall with the 18.25 Hereford to Paddington. 1 May 1988.

125. 50030 *Repulse*. Surrounded by electric multiple units, 50030 *Repulse* passes Wimbledon Traincare Depot, as it approaches Wimbledon station with the 11.15 Waterloo to Exeter St David's. The electric unit on the right, with 5848 leading, is heading for Guildford, according to the destination indicator. 1 August 1989.

126. 50031 *Hood*. Looking east from the Stoke Road bridge, *Hood* approaches Slough station with the 11.10 Newbury to Paddington. The Windsor branch diverges to the left. 21 April 1990.

127. 50031 *Hood*. Viewed from the A329 road bridge, 50031 *Hood* and 50023 *Howe* round the curve near Lower Basildon with the 09.15 Oxford to Paddington. 28 April 1990.

128. 50031 *Hood*. With Westbourne Park station in the background, 50031 *Hood* and 50023 *Howe* head for Paddington, past Subway Junction, with the 13.00 Oxford to Paddington. The Metropolitan line of the London Underground, later renamed the Hammersmith and City line, can be seen on the left, crossing under the main line. A Class 47, running from Old Oak Common traction maintenance depot (TMD) to Paddington is behind the train. 31 March 1990.

129. 50031 *Hood*. Grand Junction, east of Birmingham New Street station, is the junction of the former London and North Western Railway line to Rugby and the Midland Railway lines to Bristol and Derby. Here, 50031 *Hood* passes the junction with the 11.18 York to Paddington. 20 January 1990.

130. 50031 *Hood*. It's August Bank Holiday weekend and passengers are waiting with their luggage as *Hood* arrives at Bodmin Parkway with the 09.32 Penzance to Paddington. 25 August 1984.

131. 50031 *Hood*. Hereford signal box was built by the Railway Signal Company in 1884. Here, *Hood* runs past the box after arriving at Hereford with the 1B33 14.00 from Paddington. 24 April 1988.

132. 50032 *Courageous*. With a Class 47 heading east on the left, 50032 *Courageous* runs through Moreton cutting, near Didcot, with the 12.00 Oxford to Paddington. 10 March 1990.

133. 50032 *Courageous*. With Royal Oak London Underground station behind its train, *Courageous* approaches Paddington with the 09.15 Oxford to Paddington. 7 October 1989.

134. 50032 *Courageous*. Crossing the viaduct over the Birmingham Main Line Canal at Oxley, near Wolverhampton, *Courageous* heads the empty stock of the 09.40 Paddington to Wolverhampton towards Oxley carriage sidings. 4 March 1989.

135. 50032 *Courageous*. Stechford station is located at the junction of the lines to Birmingham New Street, Coventry and Aston. Here, *Courageous* approaches Stechford Shunt Frame box, the former Stechford No.1 signal box, with the 14.41 Birmingham New Street to Paddington. 4 March 1989.

136. 50032 *Courageous*. Running via Kidderminster because of Sunday engineering work, *Courageous* passes Langley Green with the diverted 1V84 11.21 Liverpool Lime Street to Penzance. 19 February 1984.

137. 50032 *Courageous*. Passing the Great Western Railway signal box, built in 1885, *Courageous* arrives at Ledbury with the Sundays only 16.10 Paddington to Hereford. 2 June 1985.

138. 50032 *Courageous*. Working a local train, 50032 *Courageous* arrives at Lapford with the 14.14 Barnstaple to Exmouth. The locomotive will be replaced at Exeter. After British Rail was privatised, the Exeter to Barnstaple line became known as the Tarka Line. 3 August 1990.

139. 50032 *Courageous*. Running on the former London and South Western Railway line from Cowley Bridge Junction, near Exeter St David's station, *Courageous* departs from King's Nympton with the 16.05 Exeter Central to Barnstaple. 3 August 1990.

140. 50032 *Courageous*. With the freight line to Okehampton and Meldon Quarry on the right, *Courageous* passes Yeoford with the 12.57 Exeter St David's to Barnstaple. The Okehampton line diverges at the site of Coleford Junction, near Penstone. 3 August 1990.

141. 50033 *Glorious*. Colthrop Crossing signal box at Colthrop Lane, between Thatcham and Midgham, was built to a standard Great Western Railway design. Here, 50033 *Glorious* approaches the crossing with the 12.10 Newbury to Paddington. 28 April 1990.

142. 50033 *Glorious*. With Chance Brothers' glassworks in the background, *Glorious* passes under Spon Lane road bridge, West Bromwich, with the 11.18 Wolverhampton to Paddington. 23 September 1989. *(Photo courtesy Paul Dorney)*

143. 50033 *Glorious*. The present-day station at Coventry was formally opened in 1962, replacing an earlier two-platform station. Here, *Glorious* waits for departure time with the 09.15 Paddington to Manchester Piccadilly. 11 November 1989.

144. 50033 *Glorious*. The 07.30 Penzance to Aberdeen arrives at Birmingham New Street behind 50033 *Glorious*. 22 February 1986.

145. 50033 *Glorious*. Running on the down fast line, *Glorious* departs from Slough with the 12.15 Paddington to Oxford. 17 April 1990.

146. 50033 *Glorious*. A Great Western Railway type 4b signal box stands on the up platform at Moreton-in-Marsh. This replaced an earlier box and opened around 1883. Here, *Glorious* approaches the station with the 1B58 10.50 Paddington to Worcester Shrub Hill. 15 April 1982.

147. 50033 *Glorious*. Heading the 07.30 Aberdeen to Penzance, *Glorious* approaches Trench Lane level crossing at Dunhampstead, between Stoke Works Junction and Abbotswood Junction. The crossing is now equipped with automatic barriers and the Midland Railway signal box has gone. 2 May 1985.

148. 50034 *Furious*. The present-day Taplow station was opened in 1872. Here, *Furious* runs through the station with the 10.00 Oxford to Paddington. The station footbridge was refurbished in 2006. 21 April 1990.

149. 50034 *Furious*. Built in 1838, the two brick arches of Brunel's Maidenhead railway bridge over the River Thames were, at the time of their construction, the widest and flattest in the world. Here, 50034 *Furious* crosses the bridge with the 13.15 Paddington to Oxford. 21 April 1990.

150. 50034 *Furious*. Passing Wolvercote Junction, *Furious* heads for Oxford with the 14.41 Birmingham New Street to Paddington. In the background, the line to Worcester diverges to the left. 22 April 1989.

151. **50034** *Furious*. Originally named Northolt Junction, South Ruislip station was opened in 1908, to the west of the junction of the lines from Paddington and Marylebone. It was renamed South Ruislip in 1947. Here, before the line was resignalled, *Furious* passes with the 18.12 Paddington to Banbury. The station is shared with London Underground's Central line. 1 August 1989.

152. **50034** *Furious*. Crossing over the freight line to Onllwyn and Cwmgwrach, near Neath, *Furious* heads a relief working from Swansea to Paddington. The Neath and Brecon Junction signal box was built by the Great Western Railway in 1892. 5 May 1986.

153. 50034 *Furious*. Before the semaphore signals were replaced, 50034 *Furious* heads under the signal gantry, south of Taunton station, with a train to Plymouth. 6 August 1983.

154. 50035 *Ark Royal*. With Didcot gasometer in the background, 50035 *Ark Royal* and 50031 *Hood* head east with the 09.30 Oxford to Paddington. 18 March 1990.

155. 50035 *Ark Royal*. The 5A77 11.30 Old Oak Common to Plymouth empty stock passes Didcot behind *Ark Royal*. Didcot Power Station is prominent in the background. 11 March 1990.

156. 50035 *Ark Royal*. The 10.50 Penzance to Paddington passes Burngullow Junction, near St Austell, with *Ark Royal* in charge. The freight line to Parkandillack, serving various china clay works, diverges to the right. This is the site of Burngullow station, closed in 1931. 28 August 1984.

157. 50035 *Ark Royal*. With the steel foundry and works of F.H. Lloyd & Co in the background, 50035 *Ark Royal* passes Darlaston Junction, near Bescot, with an Oxley Sidings to Birmingham New Street empty stock working. 18 March 1989. *(Photo courtesy Paul Dorney)*

158. 50035 *Ark Royal*. The 15.13 Paddington to Wolverhampton runs alongside the Birmingham Canal, near Winson Green, behind 50035 *Ark Royal*. The piers of the bridge that carried the Harborne line, closed in 1963, can be seen in the canal. 13 May 1990.

159. 50035 *Ark Royal*. Passing the bracket signal at the west end of the station, *Ark Royal* approaches West Ruislip with the 06.54 Banbury to Paddington. The station is also a terminus of the London Underground Central line. 1 August 1989.

160. 50036 *Victorious*. Paddington station is the London terminus of the former Great Western main line. It originally opened in 1838, but large parts of the main line station, designed by Isambard Kingdom Brunel, date from 1854. Here, 50036 *Victorious* stands in the station after arriving with the 11.18 from York. 4 November 1989.

161. 50036 *Victorious*. Heading northwest from Birmingham New Street, *Victorious* follows the Birmingham Canal as it heads the 15.13 Paddington to Wolverhampton near Winson Green. 1 April 1990.

162. 50036 *Victorious*. The crew of *Victorious* observe the progress of their ballasting operation on Dainton Bank, near Aller Junction. 17 July 1990.

163. 50036 *Victorious*. With its permanent way duties on Dainton Bank completed, 50036 *Victorious* rounds the curve onto the sea wall at Teignmouth, as it heads its empty ballast wagons back to Exeter. 17 July 1990.

164. 50037 *Illustrious*. The 16.10 Paddington to Hereford passes Newland East signal box, between Worcester Foregate Street and Malvern Link, behind 50037 *Illustrious*. This is the site of Newland Halt, closed in 1965. 27 April 1986.

165. 50037 *Illustrious*. In early Network SouthEast livery, 50037 *Illustrious* approaches Worcester Shrub Hill with the 1B31 13.45 Paddington to Hereford. A track panel, removed from Midland Yard, is in the foreground. 5 June 1988.

166. 50037 *Illustrious*. West Ealing station is approximately six and a half miles from Paddington, between Ealing Broadway and Hanwell. Here, *Illustrious* passes the station with the 15.07 Paddington to Oxford. 11 February 1989.

167. 50037 *Illustrious*. Passing the closed station at Seaton Junction, *Illustrious* is in charge of the 14.22 Exeter St David's to Waterloo. A branch line ran from the junction to Seaton and, in 1970, Seaton Tramway was opened for tourists on the trackbed of the old railway. 13 April 1991.

168. 50038 *Formidable*. On a clear day, 50038 *Formidable* passes Horse Cove, near Dawlish, with the 11.45 Paddington to Penzance. Dawlish can be seen in the distance. *Formidable* was withdrawn in September 1988, after less than twenty years of service. 15 September 1985.

169. 50038 *Formidable*. With Worcester Shrub Hill Station signal box on the left, *Formidable* departs from the station with the Worcester to Peterborough parcels. 6 May 1988.

170. 50039 *Implacable*. With the embankment of the Chiltern line in the background, *Implacable* passes the old Aynho for Deddington station, closed in 1964, with the 14.41 Birmingham New Street to Paddington. 11 March 1989.

171. **50039** *Implacable*. The 13.15 Oxford to Paddington approaches Didcot Parkway behind 50039 *Implacable*. 12 March 1989.

172. **50039** *Implacable*. With the Church of St Peter and St Paul on the skyline, *Implacable* passes King's Sutton with the 09.40 Paddington to Wolverhampton. 11 March 1989.

173. 50040 *Leviathan*. The 08.00 Hereford to Paddington departs from Moreton-in-Marsh behind 50040 *Leviathan*. The remains of the former Shipston-on-Stour line, closed in 1960, are on the right. 25 April 1987.

174. 50040 *Leviathan*. Heading south from Oxford, 50040 *Leviathan* and 50050 *Fearless* pass Hinksey Yard with the 12.00 Oxford to Paddington. 14 January 1989.

175. 50040 *Leviathan*. Still retaining its original station building, Charlbury station was opened in 1853 on the Oxford, Worcester and Wolverhampton Railway. Here, 50040 *Leviathan* departs from the station with the 17.02 Paddington to Hereford. 22 April 1989.

176. 50040 *Leviathan*. Passing Banbury North signal box, *Leviathan* arrives at Banbury with the 14.41 Birmingham New Street to Paddington. The box has since been demolished. 1 May 1989.

177. 50040 *Leviathan*. Crossing Moorswater Viaduct, near Liskeard, *Leviathan* heads the 09.36 Liverpool Lime Street to Penzance. The viaduct is 954 feet long and 147 feet high. Note the remaining piers of Brunel's original viaduct. 31 August 1984.

178. 50040 *Leviathan*. Class 50s receive attention at Old Oak Common TMD. The locomotives are, from left to right, 50040 *Leviathan*, 50031 *Hood*, 50036 *Victorious* and 50037 *Illustrious*. The depot at Old Oak Common opened in 1906. 3 February 1985.

179. 50041 *Bulwark*. With photographers on the down platform and passengers clearly enjoying the journey along the sea wall, 50041 *Bulwark* runs through Dawlish station with the 09.33 Plymouth to Brighton. 2 September 1989.

180. 50041 *Bulwark*. Running on the up middle road, *Bulwark* passes Winchfield, between Hook and Fleet, with the 06.42 Exeter St David's to Waterloo. 28 March 1989.

181. 50041 *Bulwark*. With shunter 08845 on the front, 50041 *Bulwark* trails at the the rear of the empty stock of the 11.59 Portsmouth Harbour to Plymouth at Fratton, as it heads towards the harbour station. 30 March 1989.

182. 50042 *Triumph*. Approaching Hartlebury station, 50042 *Triumph* heads Sunday's diverted 11.25 Liverpool Lime Street to Penzance. Hartlebury Station signal box, designed by McKenzie & Holland and opened in 1876, has since been demolished. 14 April 1985.

183. 50042 *Triumph*. The diverted 1V84 11.21 Liverpool Lime Street to Penzance passes Cradley Heath behind 50042 *Triumph*. After closure, Cradley Heath East signal box was moved to the South Devon Railway. 18 March 1984.

184. 50042 *Triumph*. Rounding the curve on the line from Stoke Works Junction, *Triumph* approaches Droitwich Spa station with a southbound ballast train. The Ruston shunter is standing at the former Underwood's coal depot, on the right. 19 February 1989.

185. 50042 *Triumph*. Spetchley Loop is located between Stoke Works Junction and Abbotswood Junction on the Birmingham to Bristol main line. Here, *Triumph* is pictured approaching the loop with a southbound ballast train. 19 February 1989.

186. 50043 *Eagle*. Departing from the former Midland Railway station at Cheltenham Spa, *Eagle* is in charge of the 1M22 11.38 Plymouth to Manchester Piccadilly. A Class 119 diesel multiple unit is in Platform 1. 5 January 1984.

187. 50043 *Eagle*. Opened by the London and South Western Railway in 1845, Guildford is on the direct line from Waterloo to Portsmouth and is the interchange station for the lines to Reading and Redhill, as well as the line to Waterloo via Cobham. Here, 50043 *Eagle* departs from Guildford with the 15.30 Portsmouth Harbour to Waterloo. 31 March 1989.

188. 50043 *Eagle*. Southampton station, rebuilt in 1967, was renamed Southampton Central in 1994. Here, *Eagle* departs from the station, seventy-nine miles from Waterloo, with the 09.40 Plymouth to Portsmouth Harbour. 30 March 1989.

189. 50044 *Exeter*. The 11.10 Waterloo to Exeter St David's departs from Andover behind 50044 *Exeter*. A military train is in the yard, in association with the MOD Ludgershall open days on the Saturday and Sunday. Ludgershall is at the end of a branch line from Andover. 23 March 1986.

190. 50044 *Exeter*. Opened in 1859, St Austell station is now a Grade II listed building. Here, *Exeter* departs from the station with the 09.36 Liverpool Lime Street to Penzance. My Vauxhall Cavalier, in which I travelled many miles to photograph railways, is in the right foreground. 29 August 1984.

191. 50045 *Achilles*. Passing the fine signal box and gantry at the west end of the station, 50045 *Achilles* departs from Newton Abbot with an unidentified westbound working. The coaches, on the right, are outside the premises of David & Charles, publishers of railway and canal books. 6 October 1984.

192. 50045 *Achilles*. Running down the branch from Newton Abbot, *Achilles* passes Hollicombe with empty stock from Plymouth Laira to Paignton. 16 July 1989.

193. 50045 *Achilles*. The 09.15 Plymouth to Brighton arrives at Fareham behind *Achilles*. 11 August 1990.

194. 50045 *Achilles*. Great Malvern station was opened by the Worcester and Hereford Railway in 1860. Here, with Malvern Girls' College in the background, *Achilles* waits to depart with the 14.00 Paddington to Hereford. 20 March 1988.

195. 50045 *Achilles*. Passing a Class 117 diesel multiple unit, *Achilles* heads away from Oxford with the 16.25 Oxford to Paddington. 16 September 1983.

196. 50046 *Ajax*. Waiting to depart from Gillingham, Dorset, 50046 *Ajax* is in charge of the 13.15 Waterloo to Exeter St David's. The signal box was opened in April 1957. 22 June 1991.

197. 50046 *Ajax*. Running through the Devon countryside near Uton, between Yeoford and Crediton, *Ajax* heads an empty ballast train to Meldon Quarry. 2 August 1990.

198. 50046 *Ajax*. Passing the site of Nine Elms locomotive depot, closed in 1967, *Ajax* heads the 11.15 Waterloo to Exeter St David's away from Vauxhall. *Ajax* was withdrawn in the following March. 3 August 1991.

199. 50046 *Ajax*. The 14.40 Newbury to Paddington departs from Newbury behind *Ajax*. A diesel multiple unit is already entering the platform in the background. 11 March 1989.

200. 50046 *Ajax*. The 13.05 Paddington to Newbury runs through Newbury Racecourse station behind 50046 *Ajax*. The racecourse is just across Racecourse Road, on the right. 11 March 1989.

201. 50046 *Ajax*. Passing the signals at High Wycombe, *Ajax* approaches the station with the 17.47 Paddington to Banbury. Note the centre-pivot signals on the left. 10 May 1989.

202. 50046 *Ajax*. The line through Pershore was singled in the early 1970s. Here, *Ajax* arrives at the station with the 16.10 Paddington to Hereford. 5 May 1985.

203. 50047 *Swiftsure*. After taking over from an electric locomotive, 50047 *Swiftsure* departs from Birmingham New Street with the 1V90 10.47 Glasgow Central to Plymouth, with a portion from Edinburgh. A Class 86 electric is in the next platform. 15 October 1983.

204. 50047 *Swiftsure*. Pictured at Birmingham New Street station, *Swiftsure* entered service in December 1968 and was the fifth member of the class to be withdrawn, in April 1988. It was scrapped at Vic Berry's scrapyard, Leicester, the following year. 15 October 1983.

205. 50047 *Swiftsure*. Torre is situated north of Torquay on the Paignton line. Here, *Swiftsure* passes with the 1A33 08.50 Paignton to Paddington. The signal box was built in 1921, replacing an earlier box, opened in 1883. 21 September 1985.

206. 50048 *Dauntless*. Passing a container terminal at Millbrook, near Southampton, *Dauntless* heads the diverted 12.40 Waterloo to Exeter St David's. 12 August 1990.

207. 50048 *Dauntless*. Taking the middle road at Surbiton, *Dauntless* passes with the 13.10 Waterloo to Exeter St David's. 15 April 1989.

208. 50048 *Dauntless*. Worting Junction is the junction of the lines to Waterloo, Salisbury and Southampton. Here, 50048 *Dauntless* passes Worting Junction and runs under Battledown Flyover, three miles west of Basingstoke, with the 16.40 Waterloo to Salisbury. The flyover carries the up line from Southampton over the Salisbury lines. 26 June 1989.

209. 50048 *Dauntless*. Running through pleasant countryside east of Warminster, *Dauntless* heads Sunday's diverted 10.55 Waterloo to Exeter St David's. 25 November 1990.

210. 50049 *Defiance*. Now preserved on the Severn Valley Railway, *Defiance* heads away from Clapham Junction with the 15.10 Waterloo to Exeter St David's. 15 April 1989.

211. 50049 *Defiance*. Leaving Dainton Tunnel, *Defiance* heads the 08.38 Manchester Piccadilly to Penzance. Dainton Tunnel, between Newton Abbot and Totnes, is at the summit of Dainton Bank. The signal box was built by the Western Region in 1965. 29 June 1985.

212. 50049 *Defiance*. Viewed from the A240 road bridge, 50049 *Defiance* passes Surbiton with the 08.17 Exeter St David's to Waterloo. 15 April 1989.

213. 50049 *Defiance*. Approaching Pinhoe station on the eastern outskirts of Exeter, 50049 *Defiance* and D400 (50050 *Fearless*, without nameplates) head the 08.10 Exeter St David's to Waterloo. Pinhoe station closed in 1966 and a new station opened in 1983. 29 June 1991.

214. 50050 *Fearless*. Entering service in October 1967, D400, later renumbered 50050, was the first member of the class to be built. It is seen here at Vauxhall, running as D400 and without nameplates, after repainting into its original livery earlier in the year. The working is the 13.15 Waterloo to Exeter St David's. 7 December 1991.

215. 50050 *Fearless*. On the last day of the Class 50s on the Waterloo to Exeter trains, D400 and 50007 *Sir Edward Elgar* pass Coker, west of Yeovil Junction, with the 09.28 Exeter St David's to Waterloo. 24 May 1992.

216. 50050 *Fearless*. Running along the sea wall, 50050 *Fearless* heads away from the Dawlish station stop and approaches Kennaway Tunnel with the 08.02 Paddington to Paignton. Kennaway is the first of five tunnels between Dawlish and Teignmouth. 28 July 1990.

217. 50050 *Fearless*. Approaching Bescot station, *Fearless* heads empty stock from Oxley to Birmingham New Street. 8 April 1989.

218. 50050 *Fearless*. Approaching the Walwyn Road bridge, *Fearless* heads away from the Colwall station stop with the 13.45 Paddington to Hereford. This was a popular location for photographs, with the house close to the railway, until vegetation obscured the view. 23 April 1989.

219. 50050 *Fearless*. The 14.00 Paddington to Hereford departs from Evesham behind *Fearless*. Across the car park are some buildings from the Midland Railway station, on the former Ashchurch to Barnt Green line. 6 March 1988.

CLASS 50s ON PRESERVED RAILWAYS

220. Running in undercoat, 50015 *Valiant* approaches Ewood Bridge, with a train from Rawtenstall to Bury Bolton Street, during an East Lancashire Railway diesel gala. This is the site of Ewood Bridge and Edenfield station, closed in 1972. The locomotive is owned by the Bury Valiant Group. 17 June 1995.

221. Leaving Greet Tunnel, near Winchcombe, privately owned 50021 *Rodney* heads for Gotherington with a train from Toddington, on the Gloucestershire Warwickshire Railway. 20 March 1994.

222. With English Electric diesels D8001 and 40012 *Aureol* on the left, 50007 *Sir Edward Elgar* stands outside the shed at Swanwick Junction, during a diesel gala at the Midland Railway, Butterley. *Sir Edward Elgar* was purchased by the Class 40 Appeal and moved to Butterley in July 1994. 27 March 1999.

223. **50007** *Sir Edward Elgar* stands outside the shed at Bridgnorth locomotive depot, during a Severn Valley Railway diesel gala. An 08 shunter is on the right. The locomotive is now owned by the Class 50 Alliance, based on the Severn Valley Railway, and once again carries its original name, *Hercules*. 30 September 2000.

224. During the 1993 Severn Valley Railway diesel gala, 50031 *Hood* crosses Oldbury Viaduct, south of Bridgnorth, with a train to Kidderminster. *Hood* was purchased by two Fifty Fund members in December 1991. It is now operated by the Class 50 Alliance. 8 May 1993.

225. In October 1993, the Paignton and Dartmouth Railway held a Class 50 day. In action were D400, 50002, 50007, 50033 and 50042. Here, D400 heads a demonstration freight train over Hookhills Viaduct towards Kingswear. The locomotive remained in main line service until March 1994. 16 October 1993.

226. In December 1991, the Devon Diesel Society purchased 50002 *Superb*, initially for use on the Paignton and Dartmouth Railway. The first runs were for members and shareholders, but although I had tickets, I elected to photograph the trains instead of riding on them. Here, *Superb* is climbing Goodrington bank, as it heads away from Paignton, with one of the special trains on that first day. 25 April 1992.

227. In 1987, 50049 *Defiance* was renumbered 50149, equipped with lower-geared bogies, and painted in grey livery for use on freight trains. By 1989, the locomotive had reverted to its previous identity of 50049. It was withdrawn in 1991 and purchased by the Class 50 Society for use on the West Somerset Railway. Here, *Defiance* is approaching Watchet, in Railfreight two-tone grey livery as 50149, with a train to Minehead during a diesel gala. The white headboard reads 'Inaugural 50149 Working', although this was not its first train of the weekend. 29 October 1994.

CLASS 50s ON RAILTOURS

228. DAA Tours' 'The South Wales Venturer' ran from London Bridge to Tondu. Here, 50040 *Leviathan* passes the panel signal box and approaches North Junction at Westbury. *Leviathan* had brought the train from London Bridge. It was replaced by 56036 at Westbury. 20 August 1983.

229. Viewed from above Rainbow Hill Tunnel, 50050 *Fearless* and 50024 *Vanguard* approach Tunnel Junction signal box as they head Pathfinder Tours' 'The Fellsman II', from Taunton to Carlisle, away from Worcester Shrub Hill station. The Class 50s worked through to Carlisle. 23 April 1988.

230. Pathfinder Tours' 'The Taw & Tor Tourer' ran from Manchester Piccadilly to Meldon Quarry. Here, 50020 *Revenge* passes Bathampton with the returning special. *Revenge* hauled the train between Bristol Temple Meads and Meldon Quarry. 5 May 1990.

231. Because 'The Taw & Tor Tourer' did not visit Barnstaple in May as advertised, Pathfinder Tours ran a second train from Manchester Piccadilly in September, 'The Taw Retour', visiting Barnstaple, Exmouth and Heathfield. Between Bristol Temple Meads and Exeter St David's, 50031 *Hood* was the motive power, while 50032 *Courageous* returned the train to Bristol. The two 50s ran top and tail on the Devon branches. Here, *Courageous* leads the train along the Heathfield branch at Newton Abbot. *Hood* is out of sight on the rear. 16 September 1990.

232. Viewed from the top of Brunel's atmospheric railway pumping station, 50031 *Hood* runs alongside the River Exe estuary, as it passes Starcross with Pathfinder Tours' 'The Taw Retour'. *Courageous* is on the rear of the train. 16 September 1990.

233. With 50031 *Hood* leading and *Courageous* on the rear of the train, Pathfinder Tours' 'The Taw Retour' approaches Lympstone as it follows the River Exe estuary towards Exmouth. 16 September 1990.

234. In May 1991, Pathfinder Tours ran 'The Cornish Centurion II' railtour from Manchester Piccadilly to Carne Point, Drinnick Mill, Penzance and Bugle, using 50008 *Thunderer* and 50015 *Valiant*. The Class 50s headed the train from Bristol Temple Meads to Cornwall and back to Bristol, running top and tail on the Cornish branches. Here, *Thunderer* passes Golant as it heads down the branch to Carne Point. *Valiant* is on the rear of the train. 4 May 1991.

235. After visiting Drinnick Mill, 50008 *Thunderer* trails at the rear of the 'The Cornish Centurion II' railtour, as it heads away from Burngullow behind 50015 *Valiant*. The Burngullow works of English China Clays Ltd is behind the train. 4 May 1991.

236. With 50015 *Valiant* trailing at the rear, Pathfinder Tours' 'The Cornish Centurion II' arrives at Penzance behind 50008 *Thunderer*. 4 May 1991.

237. Pathfinder Tours' 'The Hoover Hoop' railtour passes St Mary's Crossing, near Brimscombe, Stroud, behind 50015 *Valiant* and 50008 *Thunderer*. *Valiant* is painted in Dutch civil engineers' grey and yellow livery. The railtour ran from Paddington to Paddington via Swindon and Cardiff Central and was hauled by the two Class 50s throughout. 19 October 1991.

238. A number of special trains were run in conjunction with the Hereford Rail Festival in May 1991. Here, 50015 *Valiant* passes Stoke Edith, between Hereford and Ledbury, with the 12.40 Hereford to Worcester Shrub Hill. 'Hereford Swansong' is written on the board behind the windscreen. 5 May 1991.

239. Approaching Maiden Newton station behind 50029 *Renown* is 538 Railtours' 'South Western 16CSVT' special train, returning from Weymouth to Waterloo, via Swindon and Eastleigh. *Renown* was used throughout the tour. 26 October 1991.

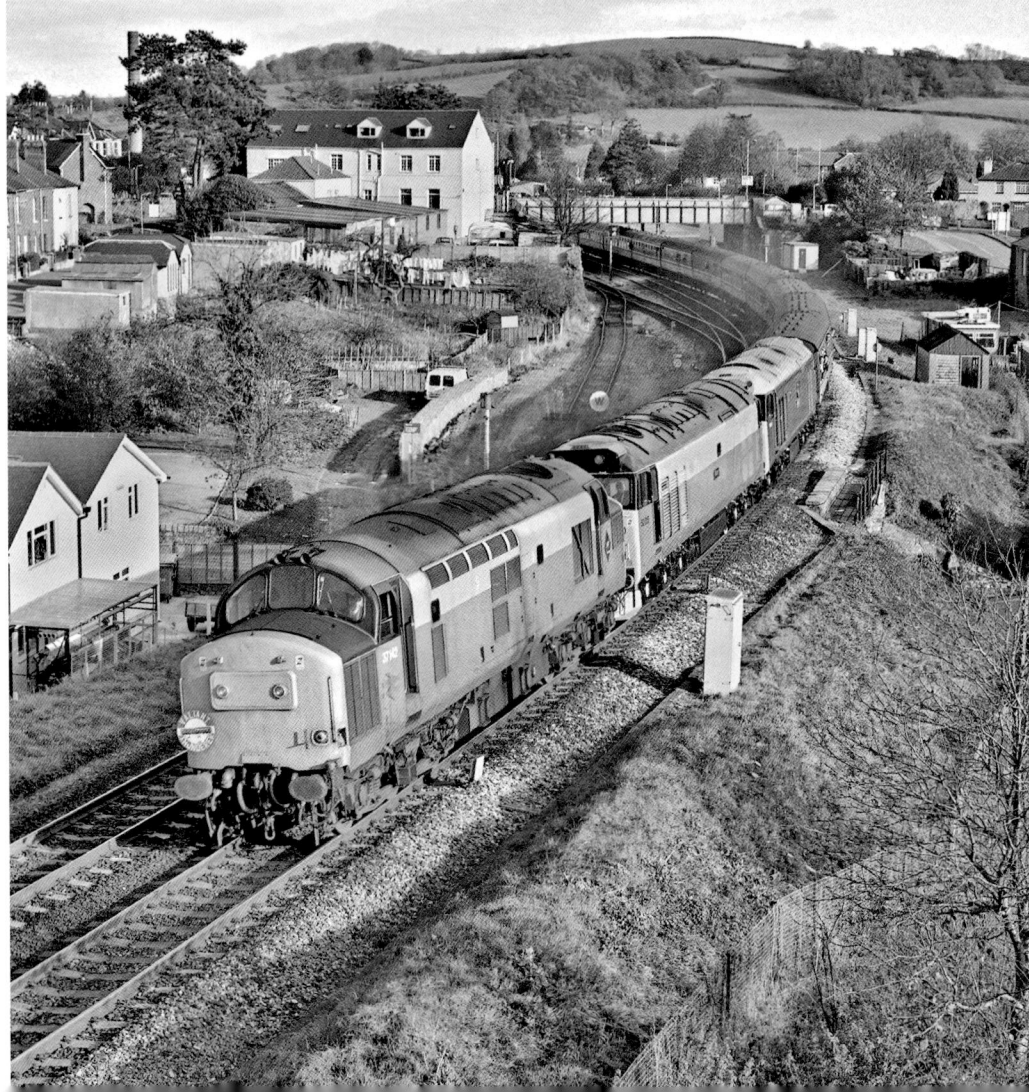

240. Pathfinder Tours' 'The Valiant Thunderer' heads away from Totnes and starts the climb of Rattery Bank behind 37142, 50015 *Valiant* and 50008 *Thunderer*. The tour ran from Manchester Piccadilly to Newquay. The Class 37 was used as pilot locomotive from Newton Abbot to Plymouth after *Valiant* developed a fault with its speedometer. 23 November 1991.

241. Visiting old haunts, 50007 *Sir Edward Elgar* and D400 pass Greenholme, on the climb to Shap Summit, with 'The Carlisle Fifty Farewell' railtour from Waterloo to Carlisle. The railtour was organised by *Rail Magazine* in conjunction with Network SouthEast. 11 April 1992.

242. Pathfinder Tours' 'The Minster Marauder', from Eastleigh to York, passes Toton behind 50007 *Sir Edward Elgar* and 50033 *Glorious*. The Class 50s were used between Bristol Temple Meads and York. Toton TMD is in the background. 7 November 1992.

243. Pathfinder Tours' 'The Hoovering Druid', from Manchester Piccadilly to Pontycymmer, Margam Abbey Works East and Ebbw Vale, passes Ebbw Junction, Newport, behind 50033 *Glorious* and D400. The Class 50s hauled the train from Derby, with 37212 attached to the other end of the train for the Welsh branches. 4 April 1992.

244. Pathfinder Tours' 'The Hoovering Druid' passes Aberbeeg behind 50033 *Glorious* and D400, as it heads for Ebbw Vale. 4 April 1992.

245. With 50033 *Glorious* and D400 in charge, 'The Hoovering Druid' railtour passes Aberthaw, as it heads towards Cardiff. The old Aberthaw High Level down station platform still remains. 4 April 1992.

246. 'The Hoovering Druid' railtour passes Llanhilleth as it heads for Ebbw Vale behind 50033 *Glorious* and D400. Class 37 37212 is on the rear of the train. Llanhilleth station closed in 1962 and a new station opened in 2008. 4 April 1992.

247. With 37212 on the front of the train, 'The Hoovering Druid' railtour passes Tondu, a junction on the line to Pontycymmer. The two Class 50s, 50033 *Glorious* and D400, are on the rear. 4 April 1992.

248. Viewed from the Westminster Road Bridge, 50007 *Sir Edward Elgar* and D400 lead Second Reach Railtours' 'The Court Chester' towards Chester station. The railtour ran from Waterloo to Chester via Worksop, Guide Bridge, Stockport and Altrincham, before returning to St Pancras. The lines on the right lead to Crewe. 13 June 1992.

249. **'The Court** Chester' heads away from Whitwell Tunnel, south of Worksop, behind D400 and 50007 *Sir Edward Elgar*. The line on the far left runs to Whitwell Quarry. 13 June 1992.

250. **Pathfinder Tours'** 'The Knighton Horse' ran from Manchester Piccadilly to Shrewsbury, via the Central Wales line. From Gloucester to Shrewsbury, and back to Gloucester, the train was headed by 50007 *Sir Edward Elgar* and D400. Here, the Class 50s are seen at Llanwrtyd Wells station, heading for Knighton. 23 January 1993.

251. Leaving Rainbow Hill Tunnel at Worcester, 50033 *Glorious* and 50007 *Sir Edward Elgar* head north with Pathfinder Tours' 'The Merseyman'. The railtour ran from Bristol Temple Meads to branches in the northwest and Seaforth docks. The Class 50s were used throughout, assisted by 56096 on the branches. 30 October 1993.

252. Pathfinder Tours' 'The Cornish Caper' approaches Keyham station, Plymouth, behind 50033 *Glorious* and 50050 *Fearless*, as they head for York. The railtour ran from York to Newquay, Penzance and St Ives. The Class 50s hauled the train from Bristol Temple Meads to Cornwall and back to York. The line on the left leads to Devonport Dockyard. 19 March 1994.

253. With BP Chemicals at Baglan Bay in the background, Pathfinder Tours' 'The Dyfed Dub-Dub' passes Briton Ferry behind D400 and 50007 *Sir Edward Elgar*. The railtour ran from Crewe to Robeston, Waterston and Fishguard Harbour. The Class 50s were used throughout with 60065 assisting on the branches. 8 January 1994.

254. Past-Time Rail's 'The Pilgrim Hoover' railtour ran from Euston to Plymouth. From Euston to Birmingham International and back, 87101 provided the motive power, with 50031 *Hood* working between Birmingham International and Plymouth. Here, *Hood*, now in private ownership, departs from Bristol Temple Meads, bound for Plymouth. Class 47 47781 *Isle of Iona* is in the background 1 November 1997.

255. In 1998, a Class 50 Railtour in South Wales, using 50031 *Hood,* was organised by the Cardiff Railway Company. Here, *Hood* heads the 1Z12 14.38 Merthyr Tydfil to Rhymney via Cardiff Central, past Pontypridd Junction signal box. The McKenzie and Holland box is located between the Treherbert and Merthyr Tydfil lines, north of the station. 9 August 1998.

256. Passing Pontlottyn, 50031 *Hood* heads for Rhymney with the 1Z12 14.38 Class 50 Railtour from Merthyr Tydfil. 9 August 1998.

257. On 16 August, the Cardiff Railway Company again ran a Class 50 Railtour in South Wales, using 50031 *Hood*. Here, *Hood* arrives at Ton Pentre, between Ystrad Rhondda and Treorchy, with the 1Z10 09.55 Rhymney to Treherbert. 16 August 1998.

258. Although not strictly a railtour, 50031 *Hood* is seen here at Bargoed, substituting for the usual Class 37, during the Wales versus Australia Rugby International at Cardiff's Millennium Stadium. The working is the 15.15 Rhymney to Cardiff Central. Class 37 37228 is on sandite duty. 23 October 1999.

259. Privately owned 50050 *Fearless* arrives at Weston-super-Mare with Nenta Railtours' 'The Star Coast to Coast Explorer' from Felixstowe to Minehead. The Class 50 failed at Hatfield Peverel, later in the tour, and was assisted by 47825. 17 July 1999.

260. Painted in maroon with gold stripes, 50017, running without nameplates, passes Leominster with the 'Northern Belle' luxury train from Manchester Victoria to Bath. The locomotive was hired to the Venice-Simplon-Orient Express company (VSOE) and repainted into this livery, reminiscent of the streamlined Duchesses of the LMSR, for working this service. The locomotive, now in original Network SouthEast livery, is currently based on the Great Central Railway at Loughborough. 19 August 2000.

CLASS 50s FOR SCRAP

261. Awaiting their fate at Booth-Roe scrapyard, Rotherham, are 47412, 50045 *Achilles* and 50026 *Indomitable*. *Indomitable*, on the right, was later saved for preservation. Six Class 50s were scrapped here, 50001, 004, 016, 020, 036, and 045. 29 February 1992.

262. The cab of an unidentified class 50 stands in Vic Berry's scrapyard, Leicester. Five Class 50s were scrapped here, 50006, 012, 014, 022, and 047. 6 August 1989.

263. *Neptune* is reduced to scrap metal at Vic Berry's, Leicester. It appears that 50006's number is being cut out, so perhaps that part of the side panel has been saved. This locomotive entered service in April 1968 and was the first to be refurbished. It was withdrawn in July 1987. 29 February 1988. *(Photo courtesy Paul Dorney)*

CLASS 50s IN COLOUR

264. Pathfinder Tours' 'The Orcadian', from Swindon to Kyle of Lochalsh, Wick and Thurso, exits Rainbow Hill Tunnel, Worcester. The motive power is provided by two preserved Class 50 diesels, both carrying Highland Stag emblems as applied to locomotives based at Inverness TMD. The lead locomotive is 50031 *Hood*, masquerading as 50028 *Tiger*, with 50049 *Defiance*, running as 50012 *Benbow*, behind. 16 June 2006.

265. Running as 50028 *Tiger*, 50031 *Hood* departs from Cardiff Central with an empty stock working to Pengam. The preserved Class 50s, 50031 and 50049, were being used by Arriva Trains Wales on certain services to Fishguard Harbour at this time. 12 August 2006.

266. The 13.35 Fishguard Harbour to Cardiff Central rounds the curve near Miskin behind 50049 *Defiance*. 15 July 2006.

267. Heading the 5Z50 Cardiff Canton to Crewe empty stock, 50031 *Hood* is pictured approaching Leominster station. The former Leominster South End signal box, built around 1875 to an LNWR and GWR joint type 1 design, is on the left. 12 January 2007.

268. PTG Railtours' 'Snowdon Ranger' from Euston to Holyhead ran on 3 and 4 September 2011. With a 'Voyager' behind 50044 *Exeter* and 57304 on the rear, the 1Z51 13.14 Llanrwst to Euston passes Baschurch as the special heads back to London. 4 September 2011.

269. Network Rail (Bristol) ran 'The Bristol Coal-Stone Haul' railtour, visiting various locations around Bristol. Motive power used was 50049 *Defiance*, 31105 and 31285. Here, *Defiance* stands in the middle road at Bristol Temple Meads, next to Platform 3, before the start of the day's proceedings. 3 June 2007.

270. With both Class 50s in BR Blue livery, 50007 *Hercules* and 50050 *Fearless* approach Banbury with Pathfinder Tours' 'The Purbeck & Bomo Explorer', the 1Z50 04.48 Derby to Swanage. The two Class 50s hauled the train between Washwood Heath and Swanage, while 66720 was the traction from Derby to Washwood Heath. 11 June 2016.

271. **GB Railfreight's** 'Dub & Grub' railtour approaches Hereford behind 50049 *Defiance* and 50007 *Hercules*. The tour ran from Birmingham International to Birmingham International via the Severn Tunnel. The train then continued to Basingstoke. 5 August 2018.

272. **GB Railfreight's** 'Dub & Grub' railtour approaches Defford, as it heads for Worcester behind 50049 *Defiance* and 50007 *Hercules*. On this side, the locomotives carried the numbers and names of two Class 50s that were withdrawn in 1987, 50011 *Centurion* and 50006 *Neptune*. 5 August 2018.

273. On a stormy day, 50044 *Exeter* and 50049 *Defiance* take the line to the Severn Tunnel, as they approach Severn Tunnel Junction station with Rail Blue Charters' 1Z49 06.04 Manchester Piccadilly to Minehead. 18 October 2008.

274. Carrying its original number D444, 50044 *Exeter* runs alongside Northwood Lane as it approaches Bewdley, on the Severn Valley Railway, with a Bridgnorth to Kidderminster train. No. 50044, withdrawn in January 1991, was purchased by The Fifty Fund for £5,044.00 plus VAT. It was repainted in two-tone green livery in 2007. 8 October 2009.

275. With Stoke Works Junction in the distance, 50008 *Thunderer* and 56104 head towards Bromsgrove with rail grinders DR79501 to DR79507. The working is the 4Z03 08.00 Okehampton to Chaddesden Sidings. *Thunderer* is owned by Garcia Hanson, based at Washwood Heath. 13 January 2018.

276. A light engine movement, running as 0Z50, consisting of privately-owned locomotives moving from Bishops Lydeard, on the West Somerset Railway, to the Midland Railway, Butterley, passes Badgeworth near Cheltenham Spa. The locomotives are 50007 *Hercules*, 50049 *Defiance* and English Electric Type 1 Bo-Bo diesel-electrics 20059 and 20188. 14 June 2017.

277. Standing next to the turntable at Tyseley Locomotive Works are 50021 *Rodney* and 50017 *Royal Oak*. *Rodney*, in Large Logo livery, was withdrawn in April 1990. *Royal Oak*, in VSOE maroon and gold livery, was withdrawn in September 1991 and is currently based on the Great Central Railway. 20 August 2006.

278. In two variations of Large Logo livery, 50030 *Repulse*, left, and 50029 *Renown* stand at Rowsley South, Peak Rail. *Repulse* was withdrawn in April 1992 and *Renown* in March of the same year. The locomotives are owned by the Renown Repulse Restoration Group. 12 May 2007.

279. In lined Brunswick Green livery, as applied in 1984, 50007 *Sir Edward Elgar* approaches Kinchley Lane bridge, with a Great Central Railway Loughborough to Leicester North service. *Sir Edward Elgar* was withdrawn in 1991 and reinstated in 1992, for railtour use. It is now based at the Severn Valley Railway, carrying its original name, *Hercules*. 11 September 2009.

280. Withdrawn in August 1990, 50035 *Ark Royal* was the first Class 50 to be preserved. It was purchased by The Fifty Fund in 1991 and based on the Severn Valley Railway. Here, painted in Loadhaul livery and renumbered 50135, *Ark Royal* departs from Kidderminster with a Severn Valley Railway service to Bridgnorth. 4 October 2012.

281. Painted in InterCity livery, 50031 *Hood* heads for Bewdley on the Severn Valley Railway, with a Bridgnorth to Kidderminster working. *Hood* was withdrawn in August 1991. It is now operated by the Class 50 Alliance, based on the Severn Valley Railway. 5 May 2018.

282. In the same location as the previous photograph, 50026 *Indomitable*, in revised Network SouthEast livery, is on a Severn Valley Railway working from Bridgnorth to Kidderminster. Withdrawn in December 1990, it is now privately owned. 26 July 2014.

283. In BR blue livery, 50035 *Ark Royal* passes Kidderminster Station signal box, opened in 1984, as it departs from Kidderminster, Severn Valley Railway, with a train to Bridgnorth. 20 August 2016.

284. Running down Eardington Bank, 50026 *Indomitable* heads a train of LNER teak coaches on a Severn Valley Railway working from Bridgnorth to Kidderminster, during the 2012 diesel gala. 6 October 2012.

285. Rounding the curve from Bury Bolton Street, 50015 *Valiant* heads for Heywood with an East Lancashire Railway service during the spring diesel gala. *Valiant*, withdrawn in June 1992, is owned by the Bury Valiant Group. 9 March 2014.

286. Visiting from the Severn Valley Railway, 50035 *Ark Royal* departs from Toddington with a train to Cheltenham Race Course, during a diesel gala at the Gloucestershire Warwickshire Railway. 13 October 2019.

287. The surface of the River Severn is smooth, producing a nice reflection of 50015 *Valiant* as it crosses Victoria Bridge with a Bridgnorth to Kidderminster train, during the Severn Valley Railway October diesel gala. *Valiant* is based on the East Lancashire Railway. 2 October 2014.

288. The Class 50 Golden Jubilee Gala was held on the Severn Valley Railway between 4 and 6 October 2018. Among the locomotives in action was 50033 *Glorious*, running in undercoat and working its first passenger trains since 2004. For a donation towards its repaint, it was possible to sign your name or write a message on the locomotive. Here, *Glorious* is seen heading north from Foley Park Tunnel. Since this was only the first day of the gala, there is not yet too much writing on the locomotive. 4 October 2018.

289. One year later and 50033 *Glorious*, now painted in Large Logo livery, is again in action during a Severn Valley Railway diesel gala. This time *Glorious* is approaching Foley Park Tunnel. The locomotive carries a headboard which reads 'Glorious Sunrise'. The headboard is similar to the 'Glorious Sunset' headboard used on an excursion hauled by *Glorious* from York to Scarborough on 20 March 1994, with 'Sunset' changed to 'Sunrise'. *Glorious* was withdrawn from service five days after the railtour. 6 October 2019.

290. Following close cooperation with GB Railfreight, in 2019 50007 *Hercules* and 50049 *Defiance* were repainted in GBRf livery. The locomotives, still based on the Severn Valley Railway, operate selected GBRf trains, including railtours, on a spot-hire basis. Here, 50007 *Hercules,* in GBRf livery, approaches Hay Bridge, near Eardington, with a Severn Valley Railway working from Kidderminster to Bridgnorth. 14 September 2019.

291. After withdrawal from service, many HST power cars were stored at the ex-MOD site at Long Marston. Here, in GBRf livery, 50007 *Hercules*, running as 50014 *Warspite*, passes Lower Moor near Pershore with HST power cars 43196, 071, 086 and 075. The working is the 5Z92 12.32 Long Marston to Burton-on-Trent Wetmore Sidings. 17 December 2020.

292. In GBRf livery, 50049 *Defiance* is on the lifting jacks at Kidderminster diesel depot on the Severn Valley Railway. Opened in 2016, the depot is used to maintain the home fleet of diesel locomotives. 17 May 2019.

293. Running as the 0Z50 09.30 Eastleigh Works GBRf to Kidderminster Severn Valley Railway, 50007 *Hercules* and 50049 *Defiance*, in GBRf livery, run through Droitwich Spa. The locomotives had previously taken 50033 *Glorious* to Eastleigh for repainting. 21 August 2019.

294. Passing some of the semaphore signals at Worcester Shrub Hill station, 50049 *Defiance* heads for Exeter St David's. The working is the 12.58 from Kidderminster Severn Valley Railway. 16 October 2019.

295. **Following the** withdrawal of many of the InterCity 125 train sets, 50007 *Hercules* trails at the rear of a rake of surplus coaches at Lower Moor, near Pershore. The working is the 5Z51 12.23 Long Marston to Newport Docks, with 50049 *Defiance* on the front of the train. 6 October 2020.

296. **A second** view of the 5Z51 12.23 Long Marston to Newport Docks. After reversing at Worcester, 50007 *Hercules* heads the train past Besford, north of Eckington. *Defiance* is out of sight on the rear of the train. 6 October 2020.

297. Approaching Quorn and Woodhouse station during the Great Central Railway's diesel gala in September 2019, 50017 *Royal Oak* hauls a demonstration freight train, made up of 16-ton mineral wagons. 6 September 2019.

298. During the Class 50 Golden Jubilee Gala in October 2018, 50008 *Thunderer* exits Foley Park Tunnel, on the Severn Valley Railway, with a demonstration freight train made up of loaded ballast wagons. The gala hosted the largest gathering of a single type of locomotive in railway preservation history, with eleven out of the eighteen preserved Class 50s attending the event. The locomotives running were 50007, 008, 015, 017, 031, 033, 035, 044, 049 and 050. Also present was 50026, as a static exhibit. 4 October 2018.

CLASS 50 NAMEPLATES AND CRESTS

299. The nameplate and crest of 50008 *Thunderer*, named after a dreadnought battleship on 1 September 1978. 5 October 2018.

300. The nameplate and crest of 50026 *Indomitable*, named after an aircraft carrier on 29 March 1978. 5 October 2018.

301. The nameplate and crest of 50031 *Hood*, named after a battlecruiser on 28 June 1978. 5 October 2018.

302. The nameplate and crest of 50049 *Defiance*, named on 2 May 1978 after the Royal Navy's torpedo school. Twelve ships and two shore establishments have been named HMS *Defiance*. 5 October 2018.

PORTUGUESE 1800 CLASS

303. English Electric Co-Co diesel-electric 1801, built at Vulcan Foundry in 1968, stands at Barreiro locomotive depot, across the Tagus from Lisbon. A second member of the class is in the background. Ten of these 1,668mm broad gauge locomotives, 1801 to 1810, were supplied to Comboios de Portugal (Portuguese Railways) in 1968 and 1969. They were based on the British Railways D400 series (Class 50), but with conventional control gear. They used the same English Electric 16CSVT as the Class 50s, but the generator and traction motors were similar to those used on the British Railways Type 3 and Deltic locomotives. All the class were withdrawn in 2001, but 1805 was preserved in operational condition. 16 March 1992.

BIBLIOGRAPHY

Baker, S.K., *Rail Atlas Great Britain and Ireland* (Haynes Publishing Group, 1988).
British Rail, *British Rail Passenger Timetable(s)*, May 1984–October 1993 (British Railways Board, 1983–1993).
British Rail Class 50, https://en.wikipedia.org/wiki/British_Rail_Class_50.
Clough, D., Beckett, M. and Hunt, M., *A Guide to Portuguese Railways* (Fearless Publications, 1991).
Jowett, A., *Jowett's Railway Atlas of Great Britain and Ireland* (Patrick Stephens Ltd, 1989).
Loader, Martin, Martin Loader's Railway Photography, www.hondawanderer.com.
Marsden, C., *35 Years of Main Line Diesel Traction* (Oxford Publishing Co., 1982).
Marsden, C., *Life & Times Series: Class 50* (Oxford Publishing Co., 1991).
Saunders, K. and Clough, D., *Class 50 Factfile* (Class 50 Society, 1993).
Vaughan J., *Profile of the Class 50s* (Oxford Publishing Co., 1983).
Wood, R., *British Rail Locomotives* (Ian Allan Ltd, 1986).

INDEX TO CLASS 50 LOCOMOTIVES BY PHOTO NUMBERS

50001, 2-6, 15
50002, 7-14, 226
50003, 15-17, 108
50004, 18-21
50005, 22-26
50006, 27-28
50007, 29-32, 82, 215, 222-223, 241-242, 248-251, 253, 263, 270-272, 276, 279, 290-291, 293, 295-296
50008, 33-37, 234-235, 237, 240, 275, 298-299
50009, 38-42
50010, 43-44
50011, 45-46
50012, 47-48
50013, 49
50014, 50-52
50015, 53-57, 220, 236-238, 240, 285, 287
50016, 58-59
50017, 60-65, 260, 277, 297
50018, 66-67
50019, 68-69
50020, 70-71, 230
50021, 72-75, 221, 277
50022, 76-78
50023, 79-85, 127-128
50024, 86-93, 229
50025, 94-96
50026, 97-103, 261, 282, 284, 300
50027, 104-107
50028, 108-114
50029, 115-119, 239, 278
50030, 120-125, 278
50031, 126-131, 154, 178, 224, 232-233, 254-258, 264-265, 267, 281, 301
50032, 132-140, 231
50033, 141-147, 242-247, 251-252, 288-289
50034, 148-153
50035, 154-159, 280, 283, 286
50036, 160-163, 178
50037, 164-167, 178
50038, 168-169
50039, 170-172
50040, 173-178, 228
50041, 179-181
50042, 2, 182-185
50043, 186-188
50044, 189-190, 268, 273-274
50045, 191-195, 261
50046, 196-202
50047, 203-205
50048, 206-209
50049, 210-213, 227, 264, 266, 269, 271-273, 276, 292-294, 302
50050, 32, 174, 213, 214-219, 225, 229, 241, 243-250, 252-253, 259, 270

INDEX TO LOCATIONS BY PHOTO NUMBERS

Abbotswood Junction, 16, 40
Aberbeeg, 244
Aberthaw, 245
Adderley Park, 99
Aldermaston, 87
Aller Junction, 28
Andover, 189
Aynho Junction, 91
Aynho, 170

Badgeworth, 276
Banbury, 176, 270
Bargoed, 258
Barnt Green, 48
Barreiro, Portugal, 301
Barry, 68
Baschurch, 268
Basingstoke, 104
Bathampton, 230
Battledown Flyover, 208
Bescot, 70, 217
Besford, 296
Birmingham University, 122
Birmingham, 129, 144, 158, 161, 203, 204
Blackwell, 2, 51, 123
Bodmin, 74, 130
Bristol, 52, 254, 269
Briton Ferry, 253
Burngullow Junction, 156, 235
Bursledon, 41

Cardiff, 265
Castle Cary, 64, 96
Chard Junction, 9
Charlbury, 175
Cheltenham Spa, 84, 186
Chester, 248
Clapham Junction, 86, 210

Cockwood Harbour, 18, 66
Coker, 215
Colthrop Crossing, 141
Colwall, 218
Coventry, 143
Cradley Heath, 183
Crediton, 53
Crewkerne, 67, 115
Culham, 80

Dainton Bank, 162
Dainton Tunnel, 211
Darlaston Junction, 157
Dawlish, 1, 14, 45, 85, 95, 179, 216
Defford, 272
Didcot, 79, 132, 154, 155, 171
Dinton, 11
Droitwich Spa, 20, 58, 184, 293
Dunhampstead, 26, 147

East Lancashire Railway, 220, 285
Eastleigh, 6, 29, 105
Ebbw Junction, 243
Evesham, 219
Exeter, 15, 49, 69
Exminster, 59
Exmouth Junction, 10

Fareham, 193
Fenny Bridges, 61
Fratton, 181

Gillingham, 196
Gloucestershire Warwickshire Railway, 221, 286
Golant, 234
Grateley, 12

Great Central Railway, 279, 297
Great Wishford, 107
Greaves Siding, 98
Greenholme, 241
Guildford, 187

Hampton-in-Arden, 88
Hanwell, 94
Harbury, 89
Hartlebury, 182
Hayes and Harlington, 90
Henwick, 25
Hereford, 131, 271
High Wycombe, 201
Hollicombe, 192
Horse Cove, 63, 168

Kenilworth, 93
Keyham, 252
Kidderminster, 71
King's Nympton, 139
King's Sutton, 172
Kingham, 21

Langley Green, 136
Lapford, 138
Leamington Spa, 97
Ledbury, 103, 124, 137
Leicester, 262, 263
Leominster, 260, 267
Lickey Incline, 47
Liskeard, 46
Little Langford, 112
Llanhilleth, 246
Llanwrtyd Wells, 250
Lower Moor, 291, 295
Lower Basildon, 127
Lympstone, 233

Maiden Newton, 239
Maidenhead, 149
Malvern, 65, 109, 194
Midland Railway, Butterley, 222
Milborne Wick, 32
Millbrook, 206
Miskin, 266
Moorswater Viaduct, 177
Moreton-in-Marsh, 78, 146, 173

Neath, 152
New Malden, 62
Newbury, 199, 200
Newland, 8, 164
Newport, 55
Newton Abbot, 50, 119, 191, 231
Norton Junction, 39, 56, 102

Oddingley, 19
Okehampton, 34, 35
Old Oak Common, 101, 178
Oxford, 174, 195
Oxley, 134

Paddington, 81, 128, 133, 160
Paignton and Dartmouth Railway, 225, 226
Par, 37
Penzance, 38, 236
Pershore, 202
Pinhoe, 213
Pirbright, 116
Plymouth, 4, 22, 43
Pontypridd, 255
Pontlottyn, 256

Redbridge, 30
Rotherham, 261
Rowsley South, 278

Salisbury, 117
Saltash, 76
Seaton Junction, 167
Severn Tunnel Junction, 273
Severn Valley Railway, 223, 274, 280-284, 287-290, 292, 298
Shaldon Bridge, 108
Sherborne, 120
Slough, 126, 145
South Ruislip, 151
Southall, 31, 82
Southampton, 17, 188
Spetchley, 110, 185
St Austell, 190
St Denys, 113
St Mary's Crossing, Brimscombe, 237
Starcross, 232
Stechford, 135
Stoke Edith, 238

Stoke Works Junction, 75, 275
Stourbridge Junction, 100
Surbiton, 207, 212

Taplow, 148
Taunton, 33, 153
Teignmouth, 24, 163
Tisbury, 23
Tiverton Junction, 60
Ton Pentre, 257
Tondu, 247
Torre, 205
Totnes, 111, 240
Toton, 242
Tyseley, 277

Uton, 197

Vauxhall, 198, 214

Warminster, 209
Waterloo, 114

Wednesbury, 7
West Bromwich, 142
West Ealing, 166
West Ruislip, 159
West Somerset Railway, 227
Westbury, 228
Weston-super-Mare, 259
Weybridge, 83
Whitchurch, 13
Whitwell Tunnel, 249
Wimbledon, 5, 118, 125
Winchfield, 180
Wolvercote Junction, 150
Worcester, 3, 27, 36, 42, 44, 57, 72, 73, 77, 92, 106, 165, 169, 229, 251, 264, 294

Yeoford, 54, 140
Yeovil Junction, 121